Division A.—Locomotive Engines.

A COMPLETE SET OF RULES AND REGULATIONS FOR THE PRACTICAL MANAGEMENT OF A LOCOMOTIVE ENGINE, AND FOR THE GUIDANCE OF ENGINE DRIVERS.

BY JOHN SEWELL, L.E.

JOHN WEALE, AUGUST 31st, 1848.

INTRODUCTION TO THE FIRST PAPER.

LOCOMOTIVE ENGINE DRIVING has now become a very important duty, requiring the utmost vigilance on the part of the driver, which renders it necessary that he should possess at all times great presence of mind and promptitude of action when on the road.

The importance, therefore, of these duties to the public requires a proper course of training and examination before any candidate becomes qualified to act as an engine-man.

For this reason there should be a regular period of apprenticeship for every engine-man, the length of such apprenticeship being regulated by his previous mechanical knowledge and experience. The speed on railways is now about 20 miles an hour greater than it was about ten years ago; and the man who could then, with little training, drive a locomotive at a speed of from 20 to 25 miles an hour, could not be prudently appointed to do so at a speed of from 40 to 45 miles an hour, now common on all lines of railway.

We therefore think that it is only a necessary precaution on the part of Railway Companies, that every engine driver in their employment should have his indentures regularly drawn up and attested in the usual way of such deeds, as they would be alike valuable to the public, to railway companies, and to engine-men, as satisfactory testimonials of experience, competency, and good conduct.

It is, however, highly creditable to engine drivers as a body, that in most cases of emergency they have proved equal to the occasion, and all competent drivers now employed should be entitled to indentures from their present employers.

The plan, therefore, on which the following Rules are drawn up, is "to begin at the beginning," and accordingly they commence with the personal and skilled qualifications and period of apprenticeship. They next embrace pay,—conditions of service,—and premiums for economy and good conduct. The latter is a prudent plan, which works well for the employers and employed. We proceed then to the engine-man's active duties, taking them in rotation as they occur when he is on duty, and including instructions for accidents, fogs, inclines, junctions, tunnels, piloting, and when in shed,—thus closing the cycle of the usual duties of an engine driver, and forming altogether a complete set of Rules, which will be instructive to the public as well as useful to those for whom they are more especially intended.

On extensive railways with numerous branches, special regulations are necessary, such as a lamp or lamps of a particular colour to distinguish particular trains by; the rate of speed; signals, &c., at particular junctions, inclines, or tunnels,—all which can be readily drawn up according to the peculiar circumstances attending each case, from the ample means in the following pages of supplying the necessary details. Many of the Rules are framed upon the suggestions arising from the actual observation of accidents caused by neglect of the precautions conveyed in them during ten years' experience, and they are consequently more comprehensive than those formerly drawn up by Charles Hutton Gregory, C.E.

No rules or instructions can supply the place of proper training, or the want of the requisite mental and personal qualifications of the man; but where these qualifications exist, a set of rules, followed up by a course of training, furnishes a candidate with the experience of others in addition to his own, thus enabling him to carry out the instructions he receives with safety and success.

CONTENTS OF THE FIRST PAPER.

MANAGEMENT

OF

THE LOCOMOTIVE ENGINE.

QUALIFICATIONS OF ENGINE DRIVERS AND FIRE-MEN.

1. Personal.—The most essential personal qualifications of an engine-man or fire-man are,—sobriety, steadiness, activity, presence of mind on emergencies, and the most constant watchfulness when on duty. These, combined with an accurate practical knowledge of the construction of a locomotive engine and its management, are essentially necessary to any candidate before he can become a good and safe engine driver. The same qualities of mind, combined with a knowledge how to shut off and put on steam, reverse the engine, know the height of the water in the boiler, how to turn on the feeds, and apply the brakes, are necessary for every candidate before he can be allowed to go with a train as fire-man.

2. Skilled.—Every engine-man and fire-man should be able to read and write, and understand the mechanism of a locomotive engine. If they have been fitters of engines, it is by some regarded as an advantage, and should be so stated in their testimonials.

3. Apprenticeship.—Every engine-man should serve a regular apprenticeship. If the candidate has been a fitter or erecter of locomotives, or a stationary engine driver of experience, he must serve as fire-man at least one year, under a steady and competent engine-man, who must certify to his steady habits and practical knowledge of the principles of the management of an engine before the candidate is qualified to become an engine driver. If he has not been a fitter, erecter, or a qualified stationary engine driver, the candidate must serve three years as a fire-man, to the satisfaction of the locomotive superintendent, and also obtain the certificate of the engine-man (as above) before he is qualified to become an engine driver.

4. Examination.—When the candidate has served his appointed time as a fire-man, he must deliver the engine-man's certificate of his knowledge, sobriety, and good conduct, to the locomotive superintendent, who will then, in presence

of the engine-man he served under, personally examine the candidate as to his knowledge and competency to manage a locomotive in cases of emergency, and otherwise of his several duties as a driver. If the candidate pass this examination satisfactorily, he will then be eligible to act as an engine-man whenever his services may be required.

5. Indentures.—The indentures, testifying to the candidate's knowledge, competency, sobriety, regular habits, and that he has passed his examination successfully, should then be signed by the locomotive superintendent, as expressive of his confidence in the general ability of the candidate to be appointed a locomotive engine driver. This indenture should be deposited with the Secretary or other officer appointed to take charge of such documents, who will restore it to the owner, with an additional testimonial of good conduct, upon his leaving the service in a proper manner.

PAY AND CONDITIONS OF SERVICE.

6. Exclusive Service.—Every man must devote himself exclusively to the Company's service, attending at such hours as may be appointed, and residing wherever he may be required.

7. Obedience.—He must obey promptly all instructions he may receive from the locomotive superintendent, or other persons placed in authority over him by the Directors, and must strictly conform to all the regulations of the Company. Every engine-man and fire-man will be under the immediate orders of the locomotive superintendent, or his foreman, when not with a train; and they must obey all their orders, in addition to the general Rules and Regulations with which they are furnished.

8. Pay and Premiums.—The pay is usually fixed per week, and for such hours, whether early or late, as the working of the line requires; the general principle being to give to every engine-man a fair day's work.

When off duty on the line, the engine-man will be employed in the engine-shed on his own engine, if it require it, or otherwise as instructed by the foreman of the engine-house. On some of the leading lines, premiums are given for good conduct, and for economizing the consumption of coke and stores, which is found to work very well, and might be usefully extended on all lines.

9. Misconduct.—He will be liable to immediate dismissal for disobedience of orders, negligence, misconduct, rudeness, incompetency, or insobriety; and to fines, for other errors, as may be determined by the Directors of the Company.

Intoxication being a grave offence, every man so offending, besides dismissal, will be liable also to fine and imprisonment by the magistrates of the district. If dismissed for any serious fault, or if quitting the Company's service without the usual notice, the whole or any part of the arrears of his pay, then due, will be forfeited, as may be determined by the Directors.

10. Deductions.—The Company reserve the right to deduct from the pay of their men such sums as may be awarded as fines for neglect of duty, and for rent, when engine-men or fire-men are their tenants.

11. Absence.—No engine-men or fire-men must absent themselves from their duty without the permission of their superior officer, as the pay of every man absent without leave, or from sickness or other cause, will be suspended until a special order is obtained in each case from the Directors.

12. Dress.—Every engine-man and fire-man to appear in white fustian clothes, which are to be neat and clean every Monday morning, or on Sundays, when they may be required to work on that day.

DUTIES OF ENGINE-MEN AND FIRE-MEN.

13. Discretion.—As every engine-man must serve an apprenticeship to engine driving, and pass his examination before he obtains charge of an engine, it would be as unnecessary as it is impossible to lay down rules for every contingency which may occur to an engine on the road, as they require that presence of mind, discretion, and cool promptitude of action from the engine-man, which can only be obtained by experience.

The following Rules and Instructions contain, however, the principal points of practical engine driving and engine-men's duties, as adopted on the leading railways of the country.

14. Attendance.—As it is an important part of an engine-man's duty to examine his engine, and see that it is in a proper state for work before leaving the shed, he should be in attendance at least one hour previous to his time of starting, as fixed by the locomotive superintendent for each man, and the fire-man at least one hour and a half before that time.

15. Examination of Engine.—The engine-man is held responsible that his engine is in good working order before it leaves the engine-house, and that his lamps are properly trimmed when he receives them, and that they are put in their places; and also that he has a sufficient supply of both coke and water in his tender. Before starting, therefore, he must carefully inspect all the connecting-rod cotters, set-pins, and split-pins of all parts of his engine; also see that the stuffing-boxes are properly packed, try his pumps and feed-pipes, as he moves out of the shed, and examine all the wheels, axles, and brakes, so far as he can do so; and if on this inspection he should discover any thing wrong, he will

at once report it to the foreman on duty, that another engine may be provided to take the train.

16. Oiling.—The engine-man must examine and fill up all the oil and tallow cups and boxes on every part of his engine before he leaves the shed, and at the same time see that the syphon cotton, or worsted, is clean, and will feed the oil, so that every working part may be regularly supplied with oil, to prevent axles, connecting-rod bearings, eccentric straps, or slide motion, from becoming heated. From the rapid movement of these parts, they require frequent attention during the trip and at stopping stations.

17. Tools.—Every engine-man must have with him at all times the following tools, for which he will be held responsible; and any person found guilty of concealing or destroying tools not under their own charge, will be fined or dismissed, as the case may require.

- 1 set of signal lamps.
- 1 gauge and 1 hand lamp.
- 1 set of screw-keys.
- 2 monkey wrenches.
- 2 clack drawers.
- 2 set-screws for slide-rods.
- 1 top and 1 bottom clack.
- 2 syphon cups.
- 1 rough and 1 smooth half-round file.
- 3 cold chisels.
- 1 padlock and chain for tools.
- 1 hand hammer.
- 1 plug hammer.
- 1 tube plug driver.
- 6 tube plugs.
- 1 pinch-bar.
- 2 tapered bars, $2'' \times 1''$ (for rails).
- 1 coke shovel.
- 2 coupling chains and hooks.
- 2 coupling links.
- 12 spare pins and cotters.
- 1 half-in. round rod, 3 ft. long (for steam chest).
- 1 screw-jack.
- 1 set of fire irons.
- 2 fire buckets.
- 1 set of oil cans.
- 1 set of oil fillers.
- 1 oiling tube.
- Some cord, flax, spun-yarn, and rope.
- Detonating signals.
- Torches for tunnels and night.

18. Tickets.—Each engine-man, before going out with his train, must procure, from his foreman's office, a train and coke and mileage ticket, to be filled up by him in the following manner: First, the exact time at which the train is started; second, the number and description of the carriages in the train; and upon arriving at the end of his trip, he must enter the time of his arrival, with any remarks he may have to make upon his trip, such as being assisted by another engine, how far assisted, and, if late, the cause of the delay.

On the coke and mileage ticket he must enter the number of trips he runs, the names of the stations between which they are run, and the number of miles; also the time his engine stands as a pilot. On this ticket must also be entered the weight of coke he receives during the day, attested by the signature of the coke-man where it is received; the engine-man, in turn, signing the coke-man's book for the quantity he received. Great attention should be given that these tickets are properly filled up, as they exhibit very accurately the performances of the engines, when carefully made up by each engine-man. Neglect of proper entries should be punished by a small fine.

19. Time Bill.—Every engine-man must take his time table with him, and regulate by it the speed of his engine as he proceeds; the great object being to keep the train going at the speed required, from which he should vary as little as possible, in order that he may arrive at all the stations as punctually to the time as is practicable. (For forms of all these Tickets see end of Rules.)

20. Whistles, when to be used.—The whistle is intended to be used as a signal of attention or of danger, and it must be sounded on approaching stations, level crossings, and tunnels; also, during a fog, snow storm, or violent rain, it must be sounded every quarter of a mile, and likewise for warning any persons who may be on the line of the approaching engine and train. On lines where only one whistle is used, three short sharp whistles, rapidly repeated, is the signal for danger; on other lines, a better description of guard's whistle is used, with a deep tone. When either of these signals is made, the guard must immediately apply his brakes, and do all he can to stop the train.

21. Water and Coke.—The engine-man must take his engine to the water crane, have his tender filled with water, and receive as much coke as will carry him to the next watering and coking station. He must not pass one of these stations without renewing his supply of both coke and water, if he requires it. When steam is blowing off, the hot water cock should be turned on to heat the water in the tender, and he must have a full boiler, and good fire, before he goes to the train.

22. Lamps.—Before dark, or in the evening before it is dusk, or in a fog, the engine-man must see that his lamps are lighted and properly fixed in their places; and if he should have to go out without a train, then he must fix a red lamp behind the tender. Should he be on the line through any unforeseen cause without his lamps, he must procure some from the first principal station he comes to.

23. Going to Train.—The engine-man having assured himself, by a careful examination, that his engine, lamps, tools, and spare stores are all in good order, and obtained his train, coke, mileage and time tickets, must cross over to the main line, in front of the train, five minutes before the time of starting, unless ordered otherwise by some superior officer.

Great caution must be used in placing the engine against the train, which should be done without moving a single carriage, in order to guard against injury to any passenger who may be in the act of stepping into a carriage at that moment.

24. Guards' Orders.—Every engine-man will be under the orders of the first guard of the train in all matters affecting the starting, stopping, or the movements of the train. In cases of accident, he must, if required, disconnect his engine, and proceed for assistance as he may be instructed by the first guard. He

must also give both his advice and assistance to the guard in every way he can in all cases of accident, and generally obey promptly all orders or signals given to him, whether by station masters or the first guards, so far as the safe and proper working of his engine will admit.

25. Engine standing.—While the engine is standing still, whether before starting, or at a station, or on the line, for however short a time, the slides must always be thrown out of gear, the steam shut off, and the tender brake screwed tight on, until the signal be given for starting. The engine-man must be very careful to start, and stop, steadily, without jerking the train, and not shut off the steam too suddenly (unless in cases of accident), so as to cause a concussion of the carriages, or waggons, to the risk of passengers, and, in cattle trains, to the injury of the animals. He must at no time leave his engine without seeing the above rules strictly complied with, and then not without placing the engine in charge of his fire-man.

26. Persons on Tender.—Except the proper engine-man and fire-man, no person is allowed on the engine or tender without the permission of the Directors or one of the superior officers of the Company, or in case of need, during the journey, by direction of the first guard of the train.

27. Starting the Train.—Every engine-man, on receiving the signal from the first guard to start the train, must sound his whistle BEFORE TURNING ON THE STEAM; then place himself by his hand gear, and very carefully and gradually open the regulator, to prevent jerking of the carriages, slipping, priming, or other injury to his engine. When the engine has got into speed, he must then regulate the travel of his slides, according to the load, and open the regulator to that point, determined by experience, where the generation and consumption of steam take place with greater effect and economy than when the regulator is full open. Nothing should induce the engine-man to use steam of a higher pressure than is allowed by the locomotive superintendent.

28. Water in the Cylinders.—The condensed steam in the cylinders must be allowed to escape by the steady and gradual opening of the regulator, and likewise of the cylinder cock, as the sudden opening of the regulator is liable to break the piston or cylinder-cover joints, when there is water in the cylinders.

29. Priming.—When priming takes place from the boiler, the steam must be partly shut off, and the fire-door opened, which will generally stop it. But if it arises from the boiler being too full, the feeds must be shut off, to reduce the water to its proper level. When the boiler is not too full, the feeds must not be shut off, in order to save the tubes and fire-box, which would thus become exposed from the rapid waste of water by the priming, if such waste was not kept up by the feeds.

30. Slipping.—When an engine begins to slip, the steam must be nearly all

shut off, and carefully regulated until slipping ceases. Engines should have two sand-boxes, which should then be opened, that equal bite may be given to both driving wheels from the sand dropping on the rails. When slipping ceases, the slides of the sand-boxes must be shut off, and the regulator gradually opened to its proper place. A third sand-box is used for backing a train with.

31. FEEDING WATER.—The supply of water to the boiler must be carefully regulated, so as to keep both feeds on at an uniform rate all the way, thereby keeping the boiler at a proper and equal level, for the more rapid and regular generation of steam. When near the end of a trip, the feeds may be gradually increased as shall be required to fill up the boiler, before arriving at the station.

The feeds should at no time be too suddenly opened, as doing so is very liable to break the clacks.

32. FIRING.—Coke must, as far as possible, be put on the fire at such places as do not require the full power of the engine. The steam must be partly shut off when coke is put on the fire, as, when the steam is full on, the strong draft carries the coke up against the tubes, and is liable to choke them up. When approaching the end of the trip, the small coke must be used, and the fire allowed to burn as low as will take the train on to the station safely.

33. ATTENTION TO SIGNALS.—Every engine-man must keep a good look-out, as he moves forward, for any signals, either from the police or from any other person, or for any indication of danger made to him, or which he may observe himself,—all which he is responsible for seeing and immediately attending to; and he must obey any signal made by a police-man, or gate-keeper, even if he should see reason to think such signal unnecessary. The lives of the passengers are intrusted to his care, and it is fully expected that he shall not only attend to every signal given him, and to all his instructions, but also that he will, on all occasions, be vigilant and cautious himself, not trusting entirely to signals for safety.

34. DESCRIPTION OF SIGNALS.—The usual signals are as follows:

Red,	as a	signal of	*Danger,*	is	*Stop.*
Green,	,,	,,	*Caution,*	,,	*Proceed slowly.*
White,	,,	,,	*All right,*	,,	*Go on.*

DETONATING SIGNALS FOR FOGGY WEATHER.

Semaphore Signals.—Posts with two moveable arms, standing right and left to the line of rails, which are raised or depressed by a police-man, to denote *danger, caution,* or *all right,* as the case may be.

LAMP SIGNALS FOR NIGHT.

SEMAPHORE.

Stop. Go slowly. Go on.

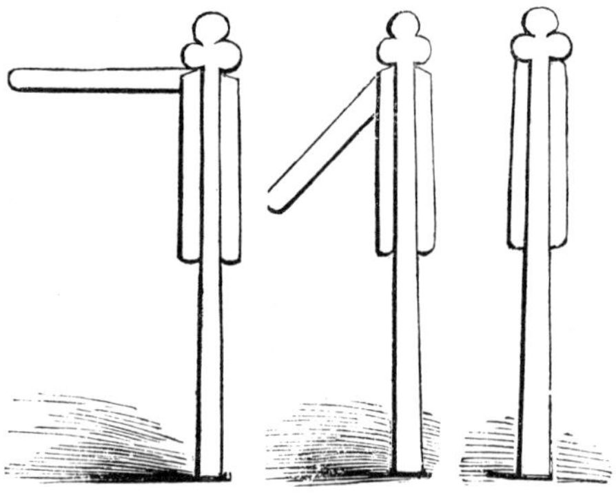

1. Danger. 2. Caution. 3. Right.

POLICE.

Go on. Go slowly. Go slowly.

4. Right. 5. Caution. Bad Road. 6. Caution. Engine ahead, on same line.

(Painted Boards.)

Stop. Go slowly.

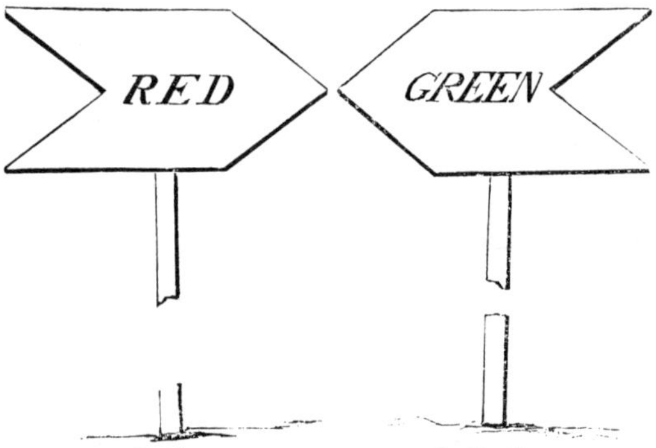

7. Danger. 8. Caution.

Go on. Go slowly. Stop.

9. Right. 10. Caution. 11. Danger.

High Masts.
Stop.

Painted Red.
12. Danger.

Go on.

Painted Red.
13. Right.

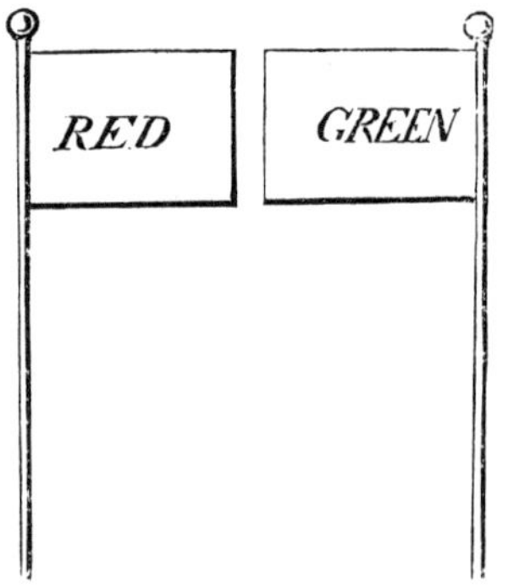

14. Hand Signals.
Danger. Caution.

Hand-flags in windy weather are a very bad signal, painted boards being much better, as they always show their full size, whilst the wind not unfrequently makes the hand-flag appear a mere line.

Waving a white light violently is also a signal of danger, when a red light is not to be had. Fog, night, and tunnel signals are made by lamps and detonators. Day signals by police-men, hand-flags, painted boards, and high-mast signals. As signals vary on different lines, each engine-man is supplied with a set of the Rules and Regulations of the Company he is employed by, describing the particular signals used, and which must be strictly attended to on every occasion.

35. Fire-men's Duties.—Fire-men are to be entirely under the orders of the engine-men while on duty. If the engine-man is engaged with any part of the engine, the fire-man must keep on the look-out, and act for the engine-man, and they must never both leave the engine at the same time. The fire-man must at all times stand up and keep a good look-out, when not otherwise engaged in attending to his other duties; and he must, each mile, look back along both sides of the train, to see that all is right, and that the guards are not signalling by hand or lamp to the engine-man. When the fire-man is so engaged that he cannot look back, he must warn the engine-man, who will then do so himself. The fire-man must also be ready, at all times, at a signal from the engine-man, to go to the brake, and, when approaching any station where the engine is to stop, or observing any obstruction on the line, or seeing any signal intimating danger or caution, it is the duty of the fire-man, without waiting for orders, to go to the brake, and to warn the engine-man instantly. Should the engine-man be rendered incapable of doing his duty, whether by accident or otherwise, the fire-man is to take the management of the engine until another driver can be obtained, proceeding with caution, and reporting the circumstance to the first guard at the earliest opportunity.

36. Guards' Signals.—Guards should regularly signal to the engine-man that all is right; but if any thing is wrong, they must give the caution or danger signal. The usual signal from the engine-man to the guards is by three short sharp whistles on some lines; on others, by a deep-toned guard's whistle,—an excellent plan, preventing the liability of any mistake whatever in the use of the whistles. On this signal being given, the guard must instantly signal that it is understood, to the fire-man, who must look back immediately the signal is given by the engine-man, to observe which signal the guard gives. This signal and answer must be made within the first half mile after starting with a goods train, and every two miles afterwards by the guards on the journey. With passenger trains, this signal must be made every mile on the journey; the whistle being used only in cases of need.

37. Engine-men's Signals.—Engine-men must also signal one another when on the road, by standing on the right-hand side of the engine, so as to be next each other in passing, and must always do so with the same signals as those used on the line, to indicate whether the line they have passed is clear, or whether there is a train ahead, or any other cause of danger existing.

38. All Trains to be Drawn.—No engine must pass along the wrong line of road, nor be allowed to propel a train of carriages or waggons; but in all cases it must draw the train after it on the right road, except in case of an engine being disabled on the road, when the succeeding engine may propel the train slowly as far as the first 'shunt' or siding, at which place the said propelling engine shall move past, and take the lead. In case of the road becoming stopped, when it will be unavoidably necessary for an engine to go back on the wrong line, the engine-man must send his fire-man, or some other competent person, back a distance of not less than half a mile before his engine moves, to warn any engine coming in the opposite direction; and the person so sent back must continue to preserve the distance of half a mile between him and the engine, until it gets into the right line again.

39. Distance between Engines.—All engines travelling on the same line must be kept at least two miles apart, unless expressly required to join on to the preceding train or engine, which the engine-man shall approach with the utmost caution. Such a step can only be justifiable under a pressing necessity.

40. Two Engines working together.—When two engines are working together, the second engine-man must watch for and take his signals from the leading engine-man; but should the second engine-man discover any thing wrong with the train, he must blow his whistle to warn the first engine-man, that the two engines may always check and stop together. Engine-men with trains requiring assistance, must in all cases, with passenger trains, allow the assistant engine to go in front, and also with goods trains, where practicable.

41. Part of Train detached.—When any part of a train is detached, while in motion, care must be taken not to stop the engine or train in front before the detached part is stopped, no matter whether it has become detached intentionally or by accident, to prevent a dangerous collision with the carriages in front, should the latter stop first. Whenever an engine-man finds his engine disconnected from the train by accident or otherwise, he must keep his engine ahead, clear of the train altogether, and in no case pull up his engine until the train has been stopped by the aid of the guards' brakes, when he can move backwards, and couple on to the train in the usual way.

It is the duty of the guard of any intentionally detached part of the train to apply the brake so as to stop at the proper place.

42. Passing Stations.—Every engine-man must be careful when he passes a station, or when the road is under repair, to proceed slowly and cautiously; and he must also do so whenever he sees the *green signal.* He must on no account pass a *red signal,* or any other which he understands to be a signal to stop, but bring his engine to a stand close to that signal.

43. Stopping at the proper Station.—No engine or train of any sort must stop at any but the appointed stations, except only when a signal is given; or in case of accident to any part of the engine or train; or when, in the judgment of the engine-man, it is necessary to prevent accident or collision.

44. Caution in stopping Trains.—Engine-men, in bringing up their trains, must pay particular attention to the state of the weather, the condition of the rails, the length of the train, and whether on a level, an ascending or descending incline, in determining when to shut off the steam, so as to reduce the speed in proper time, and enable them to have the train so completely under their command as to stop altogether, if necessary, before reaching the platform. Stations must not be entered so rapidly as to require a violent application of the brakes, and the use of the carriage brakes must be avoided as much as possible.

45. Running on Main Line.—No engine must ever be moved from any of the stations on to the main line, except when the engine-man is proceeding, at the proper time, to take his place in front of the train. When on the main line, he must never run beyond the limits fixed at each station, without a regular order, filled up and signed by the proper foreman; and he must strictly follow the instructions contained in such order, both as regards the time of starting, and the place and time of returning.

46. Standing on Main Line.—No engine must be permitted to stand on the main line (except under very special circumstances) when not attached to a train; and no engine-man must leave his engine or train, or any part thereof, on the main line, unless there be a competent person in charge to make the necessary signals: neither must any engine cross the line of railway at a station without permission, or run tender foremost, without the orders of the locomotive superintendent, or from unavoidable necessity.

EXTRA TRAINS.

47. Signal for extra Train.—Whenever a red board, or flag, by day, or an extra red lamp by night, is carried on the last carriage or waggon of a train, it is an indication that a special or extra train is to follow, that the road and stations may be kept clear for it.

48. On Single Line.—To avoid risk of collision on single lines, from the

meeting of another engine, no extra engine, with or without a train, must be allowed to pass along the line without previous notice.

LUGGAGE TRAINS.

49. Approaching Stations.—Engine-men with luggage trains must approach all stopping places at a speed not exceeding 12 to 15 miles an hour when within a quarter of a mile of the stopping place; and, when necessary, they must signal the guards or brakes-men before they apply the tender brakes.

50. Uncovered Waggons.—Engine-men must refuse to take up luggage waggons if they contain any goods of a nature to take fire from a spark falling amongst them, unless they are properly covered and tied down.

51. Goods Trains into Sidings.—Engine-men of luggage trains must not pass any siding or crossing when there is a passenger train coming in the same direction, due within fifteen minutes, but must remove from the main line to the siding; and generally all luggage, coal, and ballast trains must give way to passenger trains by going into the nearest siding.

REPAIRING ROAD.

52. Ballasting.—When any ballast train shall stop on the main line to load or unload ballast, or other materials, the engine-man must send a ballast-man back, at least a mile, with a red signal flag, or board, by day, or a red lamp by night; and the man must stand there, on the look-out, until the ballast train has moved off ten minutes, and stop any coming engine or train, and inform them of the position of the ballast train. No ballasting permitted in foggy weather, unless by a special order, or under the most urgent circumstances.

53. Caution for Repairs.—When repairs of the road are going forward, and persons employed on the permanent way have the use of the road, a signal of danger may be given by those persons, either by a red flag or a red lamp; and on observing this signal the engine-man must immediately stop the train.

54. Train of Empties.—Engines running alone, or with a train of empty carriages or waggons, must not exceed a speed of 25 miles an hour without special orders in each case, or from urgent necessity.

ACCIDENTS.

55. Stoppages.—In case of any accident to the engine or train, causing a complete stoppage, the engine-man, after giving such directions to his fire-man to open the fire-door, rake out the fire, or otherwise, as may be necessary for the

safety of the engine, must immediately seek the first guard of the train, and communicate with him, and receive his directions; and in the absence of the first guard, the engine-man must himself ascertain whether the engine and train are clear of the opposite line, and of any train passing upon it, and, if they do not appear quite clear, remove the passengers from the carriages. He must also send back a guard or his fire-man, or a special messenger, to the next police-man, to stop any trains coming up. If dusk or dark, he must see that the carriage lamps in front, and his engine lamp, all show red lights forward, and the tail lamps, as usual, show a red light backwards.

56. Hand-Pump.—Whenever, through any accident, an engine has to stand with steam up, the hand-pump must be worked to keep up the water in the boiler, the fire-door opened, the damper shut, and other means used to prevent, as far as practicable, the generation of steam.

57. Detonating Signals.—Every engine-man should be supplied with detonating signals, which he is responsible to have always ready for use when on duty. When a stoppage occurs in foggy weather, or at night, one of these signals must be placed on the rails, every 200 yards, until the guard, fire-man, or messenger has gone back a mile at least; and at the end of that distance, two of them must be placed on the rails. When any such accident happens in a tunnel, the engine-man should, if required, occasionally hold down the steam-valve, to prevent the noise, and allow orders to be given and heard more readily. If dark, and without a red lamp or detonating signals, the man sent back must make a signal by waving his light up and down violently. If by day, he must signal with the red flag, or board, or hands, and must in all cases remain until relieved by a police-man, or the obstruction has been removed.

58. Both Lines Stopped.—Should the accident stop both lines of rails, then the same precautions must be taken to place the signals on the opposite line of rail to that which the train is on, but in a contrary direction, so that both lines of rails may be protected for at least a mile from the place of obstruction.

59. Train on Fire.—Should fire be discovered in the train, the steam must instantly be shut off, and the train brought to a stand. The signals of obstruction or stoppage must then be made to protect the train, and the burning waggon, or whatever it may be, detached with as little delay as possible. Every means must then be used by the guards and engine-man to put out the fire by water from the tender, or by other means, as the necessity of the case may require. No attempt must be made to reach the nearest water crane, if it is more than 800 yards from the place where the fire is discovered, as such a course is likely to increase the damage.

60. Burst Tube.—If a tube bursts badly, the steam and feeds must both be shut off, and the engine stopped as quickly as possible. A tube-plug must then be

driven into each end of it, and the boiler filled by the hand-pump. As soon as water is seen in the glass, the fire can be got up again, and the hot fire-box will assist in doing so more quickly. As soon as the steam is up, the engine will be enabled to move on again. In most cases it will not be necessary to stop the engine, as the fire-box end can be plugged up when running, and the other end can be plugged at the first stopping station, without delaying the train.

61. Road Obscured.—When the road is obscured by steam (from a burst tube or any other cause), no approaching engine must pass through the steam or smoke until the engine-man has ascertained that the road is clear. If any engine-man see a train stopped or stopping from accident, or other cause, on the opposite line of railway, he must immediately slacken his speed, so that he may pass such train slowly, and stop altogether, if necessary. He must then ascertain the cause of the stoppage, and report it to the next station. He must also, if necessary, stop all the trains between the spot and the next station, and caution the engine-men of the stoppage. He must also render every assistance in his power in all cases of necessity and of difficulty.

62. Broken Clacks.—Should a pump or feed-pipe give way, one pump is generally sufficient to keep up the boiler; but should that one fail also, the engine must be stopped, and the clacks examined. The engine-man will know by his pet-cocks whether it is a top, middle, or bottom clack that is faulty. If a bottom or middle clack, the nut must be unscrewed, the broken clack taken out, one of his spare clacks put in, and the nut screwed on, when he will be ready to start his engine again. If it is the top clack, it can only be repaired when the steam is not up; but unless the middle clack is also bad, the pumps will still work to keep up the water in the boiler.

63. Broken Spring.—When a spring breaks, it will seldom be necessary to stop the engine until it reaches a station, when the side can be raised by moving the other wheels of the engine on to the two wedge-shaped bars, to be laid flat on the rails, which takes the weight off the wheel where the broken spring is, and allows it to be blocked up by one of the wooden wedges: the engine can then be moved off the bars, and proceed with the train. The same operation may be done by the screw-jack, but the first causes least labour.

64. Broken Machinery.—A railway locomotive is made up of two complete engines, working separately, but both attached to a double crank-axle; therefore, whenever a cylinder-cover, piston, connecting-rod, eccentric-rod, quadrant, slides, or any other working part of the engines gives way on the road, which stops the working of one engine, the engine-man must immediately stop, and remove the broken parts, or rods, so as to leave the other engine clear of all obstruction. He must then disconnect the slide of the broken engine, and adjust it so as to cover

both the steam-ports, where he must secure it by tying, if he has no other resource at hand. The best way of fixing the slide is by a set-bolt, fitted into the outside of the slide-rod guide, being screwed tight up against the slide-rod. A set-pin can also be used when a syphon-cup is screwed on the guide, by taking off the syphon, and introducing a screw of the same thread. When this is done, the engine will be able to proceed, working one cylinder only. Should the boiler have been pierced by any of the broken rods, the fire must be immediately drawn, to save the fire-box and tubes, and a competent person sent for the requisite assistance to remove the engine and train.

65. Working with one Engine.—Whenever the engine-man has only one engine in working order, he will still be enabled to proceed with a light or ordinary train to the first pilot station, where the pilot engine will proceed with the train. Should the train be too heavy, he must consult the first guard whether to proceed with part of it, or not, for assistance, and be guided by his instructions.

66. Broken Slide.—When the slide or slide-rod breaks within the steam chest, the engine must be stopped. If the slide is on the front side of the steam-ports, and the exhaust-port open, it should be pushed back through the front slide-rod gland, so far as to cover the ports. When there is no front slide-rod, but a set-bolt in the steam chest cover, by taking out the bolt the slide can be moved through that hole also; and if the slide is on the back, and the exhaust-port open from the front of the slide, it should be pushed forward through the slide-rod gland until it covers both ports. When impracticable to cover all the ports, from a bad regulator, or other cause, or keep the slide over the ports, the connecting-rod must be taken off, and the piston blocked, to prevent its moving by the pressure of the steam against it; when the engine-man can proceed with one engine, as before.

67. Hot Connecting-rod Bearing.—When the large end of the connecting-rod becomes very hot, the bearing and the crank then adhere to each other at the surface, and at each turn of the crank it tears away the solid metal, to the great danger of breaking the straps, keys, or even the rod itself, and thus leading to most extensive damage to the slide motion, piston, and cylinder, if not to the boiler also. When it cannot be cooled while running, by throwing water upon it, and using oil and tallow to prevent adhesion of the metals, the engine must be stopped, the rod taken off, the slide fixed over the ports, and the engine-man can then proceed, according to circumstances, with one cylinder, as before, until relieved by the first pilot engine. In most cases the engine-man, by reducing his speed, and carefully cooling and oiling the hot bearing, will be enabled to proceed until he comes to a stopping station. If by this heating, or by any other means, an eccentric becomes loose, it must be shifted back to its place as marked on the axle, and screwed fast again.

68. Hot Axle Bearing.—When an axle bearing becomes very hot, the train must be stopped, and the box cooled by water from the tender. The oil or grease-holes must then be cleared, and the box filled up with tallow and oil, which, by renewing at each stopping station, will usually carry the engine, tender, or carriage, safely to the end of the trip. However, should the means used to cool the bearing not succeed, the engine-man must stop at the first pilot station, and allow the pilot engine to take his train.

In most cases it will not even be necessary to stop the train until it reaches a station, when the bearing can be cooled, and the box filled as above.

69. Broken Leading Axle.—When a leading axle breaks, the engine must be stopped, the fire drawn, and assistance obtained to clear the road and forward the train. If the engine is off the rails, the first object is to clear the way for the train to proceed, the engine being got on afterwards.

70. Going off the Rails.—When both engine and tender are thrown off the rails, it is a work of time to get them on again; and the fire must be dropped, and water run off to lighten the engine and save the boiler. Assistance must then be obtained to clear the line, if obstructed, and to forward the train to its destination. After this has been done, the engine and tender must be separated, and got on to the rails separately, by raising them, by means of screw-jacks, until a temporary railway can be laid from under the wheels on to the main line. By this means they may be gradually drawn or pushed on to the road again, according to the position they are placed in.

71. Broken Crank.—When a crank-axle breaks very badly, the engine must be stopped, the connecting-rod taken off, and the slide fixed over the ports, (as Rule 64.) If it is broken at the side of the bearing for the connecting-rod, and the stay bearings are good, the engine-man will still be able to go on with the train, if a light one; or seek assistance, if it is a heavy one. If the crank is so broken that the other engine cannot work, then the fire must be drawn, and assistance obtained in the usual way.

72. Broken Trailing Axle.—When a trailing or hind axle breaks, the engine must be stopped, and the springs and axle-boxes blocked up, so as to keep the wheels as upright as possible. The ends of the axle must then be suspended at a proper height to the foot-plate by a rope, when the engine can proceed with the train, cautiously, until relieved by a pilot engine.

73. Broken Wheel or Tire.—When a wheel or tire breaks, the engine must be stopped, and assistance obtained to forward the train. If it is a driving wheel tire, the wheel must be lifted up by the screw-jack, and blocked up to the proper level, when the engine will be able to move itself, if not the train, out of the way. If it is a leading wheel, or tire of a four or six-wheeled engine, it may

probably be thrown off the rails, and, after the train has been forwarded, the engine must be got on to the rails again, as described in Rule 70.

74. Apprehension of Danger.—In case of the stoppage of either line of rail from any cause, or apprehension of danger, whether in foggy weather or otherwise, the police-man on duty must place a detonator on the line or lines of rails so obstructed, every 200 yards from the point of danger, until they are protected for at least a mile.

75. Caution to Stoppage Signals.—In all these, or any other accident causing a stoppage, due care must be taken that the signals for a stoppage are immediately attended to, in order to prevent any following train coming into collision with the one which has stopped.

FOGS.

76. Caution at Stations.—During a fog, when an engine or train stops at a station, some person must be sent back half a mile, and place a detonating signal on the rail, to stop any engine or train coming up, until the other has started from the station. Detonating signals must also be used in a similar manner whenever any engine or train is following too closely upon another engine or train, or in any case of emergency or of danger.

77. Attention to Detonators.—Whenever an engine passes over one of these signals it explodes with a loud report, and the engine-man must then immediately stop the train. The guards of that train must likewise protect it, by sending back and placing these signals every 200 yards, as before, until they have extended a mile from the train.

78. Removal of Detonators.—After the obstruction is removed, the police-man, guard, or fire-man, must remove all the signals from the rails before proceeding.

79. Distance between Engines.—No engine or train must leave a station, during a fog, less than ten minutes after any preceding engine or train; and the police-man on duty must give the engine-man the exact time when the train started, and where it is next to stop.

80. Caution for Signals.—Engine-men must always exercise great caution in foggy weather, and especially in approaching stations, from the difficulty of discerning the regular signals until close upon them; and they must be prepared to bring their engines to a stand before reaching the signal, whenever it is required.

DESCENDING INCLINES.

81. Attention to Brakes.—In descending inclined planes, engine-men, firemen, guards, and brakes-men, must take care that they have complete control over the speed of the trains, by having the brakes screwed up so far as to be able, by a single turn of the handle, to apply them forcibly to the wheels, when required to do so. The engine-man must not, however, place too much dependence upon the assistance he may get from the carriage brakes, but keep the train perfectly under his own control by shutting off the steam in time.

82. Speed.—No engine or train should descend a steep incline at a greater speed than from 20 to 30 miles an hour, unless by special orders. With very heavy trains, the speed should not exceed 20 to 25 miles an hour; and no attempt should be made to make up lost time in going down an incline.

The usual speeds down inclines from 1 in 80 to 1 in 100, are from 20 to 30 miles an hour, according to circumstances. On inclines from 1 in 37 to 1 in 80, they vary from 10 to 20 miles an hour. At the Wapping Tunnel, Liverpool, the speed is only 4 to 5 miles an hour.

ASCENDING INCLINES.

83. Dividing Trains.—When inclines are so severe as to require the train to be divided, and taken up at twice, it must be done in the following manner: The part of the train to be left must be pushed into the siding at the bottom of the incline, clear of the main line; the first part must then be taken up the incline, and placed in the siding at the top. The engine or engines will then cross over to return on the proper line for the second part of the train. When this is taken to the top of the incline, it must be left on the main line, clear of the siding, and the first part of the train taken out and coupled to it: by this means the waggons will be again in their proper positions in the train.

Great care must be taken that the last portion of the train is not allowed to run down the incline after the engine is uncoupled from it.

84. Overtaking divided Trains.—In the event of any waggons being left upon the main line at the foot of any incline, and a succeeding engine coming up, such engine must not commence propelling or drawing the said waggons until the engine which left them shall have returned, to take them away.

85. Returning on right Road.—The assistant engine must invariably return down the proper line, and must always stand at the proper place, ready with the tender well supplied with coke and water, having the steam up, and the front of the engine towards the incline.

86. Assisting Passenger Trains.—When a passenger train requires assistance up an incline, the train must be stopped, and the assistant engine coupled on in front of the train engine, and both work together up the incline; and when fairly clear of the incline, the assistant engine must be uncoupled from the train engine, and proceed ahead to the first crossing, where it will pass over to return on the right line to its station at the bottom of the incline.

87. Assisting Goods Trains.—When a goods train requires assistance, it is usually given behind, as it takes a great part of the weight off the front waggon drag-chains; and, in the event of any of the drag-chains giving way, prevents these waggons from descending the line, which they would do, without any check or control, if the engine were ahead.

Great care must be taken by the assistant engine-man behind, to prevent any over-pushing of his engine, in case any waggon break down between him and the other engine. It is best when all assistance can be given ahead of the train.

88. Caution to Pilot-man.—In moving out of the siding to assist a goods train up an incline, the pilot engine-man must take great care that the train is past, before he comes near the main line, so as not to strike the train sideways before it passes clear of the pilot engine. As soon as it is past he must move out, and approach the end of the train with great caution, so as not to run against it with any force, and must give instant attention to any signal from the leading engine-man.

89. Responsibility of Engine-man.—No engine-man should attempt to ascend, without assistance, any incline with a greater load than his engine is quite capable of taking up with certainty; and as this varies very much with the state of the weather, the gradients, and the rails, the engine-man must decide for himself, in each case, whether he requires assistance or not, and act accordingly.

EXTRA INCLINES.

90. Brakes-men.—When trains arrive at very steep inclines, either in or out of tunnels, such as those on the London and North Western Railway, where stationary power is used to drag them up the incline, or such as the Lickey incline on the Birmingham and Bristol Railway, they are placed under the charge of special brakes-men, who have instructions, in each case, and which they must strictly attend to, both as to the speed, number of carriages or waggons, according to the nature of the traffic to be taken up or down these inclines.

91. General Regulation.—As a general rule, no train is allowed to go

up or down without one of these brakes-men, whose duty it is to examine all the brakes, and satisfy himself that they are in proper order, with sufficient men to work them, before he starts a train either to go up or down one of these inclines. The speed is usually limited from 4 to 10 miles an hour, according to circumstances, and the brakes-men must at all times, whether going up or down, be sure that they have the train under perfect control with the brakes, in the event of any thing giving way.

TUNNELS.

92. Approached cautiously.—All tunnels must be approached cautiously; so that if a signal is made to stop, the train may be stopped before entering the tunnel. Every engine-man must use his whistle before entering and while passing through a tunnel, to give proper notice to the police-man at the entrance, and to any man who may be employed in the tunnel.

93. Stoppage in Tunnels.—When an engine or train breaks down in a tunnel, the torches must be lighted to ascertain the nature and extent of the accident, which must then be remedied according to instructions in preceding Rules. Great care must be taken to protect the line both ways, to prevent collision and injury to the men at work, until the line is cleared and the train forwarded.

JUNCTIONS.

94. Signals.—In approaching junctions, every engine-man must blow the whistle at least a mile from the junction, and continue to do so until the police-man in charge of the junction points gives the proper signal that the main line is clear, when the engine-man can proceed on to it with the engine or train at a speed not exceeding from 5 to 8 miles an hour. If, however, the signal be given that the main line is obstructed, then the engine-man must immediately stop the engine or train until the obstruction be removed, and the 'all right' signal be shown, when he can proceed slowly on to the main line, as before.

95. Approached cautiously.—Every engine or train must approach the junction of two lines with great caution, and at a very slow speed, so as to stop altogether before reaching the junction, if necessary. In foggy weather the engine-man must bring the engine or train to a stand before arriving at the junction, and not go on to the main line until he has learned from the police-man on duty how long the preceding train has passed, and he must stop or proceed, according to the information he receives.

REPORTS.

96. State of Engine, &c.—The engine-man must examine his engine at the end of his trip, in the same manner as he did before starting, and report to the foreman on duty, or his clerk, any thing he may observe or know to be wrong, and enter the particulars in the report-book kept for that purpose. He must also report every unusual circumstance which may have taken place on the journey, in the same manner.

97. Spare Stores replaced.—Every engine-man must have replaced, before starting with his train, any spare stores he may have used, and he must have any broken tools made good again. He must also see that his lamps are taken to the appointed place to be cleaned and trimmed again.

DUTIES IN SHED.

98. Regulation of Weight.—The weight of every engine should be properly distributed on the wheels and springs. No engine can run steadily with one, two, or three tons more weight upon one spring than upon the opposite one. A six-lever weighing machine should be on all extensive railways for this purpose, by which the weight on each wheel can be regulated with great nicety. For passenger train engines, there should be nearly as much weight on the leading as on the driving wheels. Eight-wheeled engines have more weight on the two leading than on the driving wheels. All the weight required on the driving wheels is as much as to produce sufficient adhesion, in ordinary cases, to draw the train, as there are times when all the weight would not prevent slipping. For all ordinary trains, the engine will run the whole distance in less time and more steadily with lightly loaded driving wheels; for any time lost in starting is more than made up afterwards by the ease given to the machinery. Four and six-wheeled coupled engines have generally all the weight brought into action for power, and not for speed; and it is therefore more equally distributed over the wheels.

99. Hand-Pump to be tried.—Every engine-man, when in the shed, should try his hand-pump, to insure himself that it is in good order, as, being seldom required, they are liable to become 'furred up' with the deposit from the water, and in that case of no service in any emergency.

100. Cleaning Boiler.—The boiler should be regularly blown out, over the pit provided for that purpose, at such times as the state of the water requires, which the engine-man can readily determine from experience. The boiler must also be washed out with cold water, at least once a week, to remove all loose

deposit from it. Muriate of Ammonia is now used for preventing the adhesion of deposit to the boiler, which it effects by keeping the lime in solution, and which is blown out two or three times a week, according to the purity of the water used for the generation of the steam.

101. Adjustment of Machinery.—Every engine-man, when in the shed, must carefully examine his engine; and, when necessary, adjust the pistons, slides, connecting-rod bearings, axle bearings, springs, or any other part requiring adjustment, and see that the eccentric sheaves are all fast in their right places.

102. Trimming Stuffing-Boxes and Syphons.—He must also pack his stuffing-boxes, examine his clacks and feed-pipes, clean and trim his syphons, so that he may have his engine in good working order when required to go on duty again.

PILOTING.

103. Tools.—Pilot engine-men must at all times during the hours appointed for them to pilot, stand ready to start at once, with their steam up, their tender full of coke and water, and be provided with the following tools; namely, two pinch-bars, two screw-jacks, a large drag-chain, a rope, two signal lamps, two hand lamps, and detonators. One of the signal lamps to be put on the front of the engine, and the other behind the tender; and the hand lamps to be used as signals, or as may be required.

104. Orders.—When it is necessary to send the pilot engine to look for a train, a regular order must be made out for it; and if at or near to night, both lamps must be lighted, and a red light shown behind, and a green one in front. When the pilot engine-man meets with the train, if it be at a stand, he must ascertain the cause, show red lights both ways, and run on to the next crossing ahead. He must then cross, returning to give every assistance he can, whether the engine be broken down, or any part of the train be off the line, or to push the train slowly before him until he can put it into a siding, and get in front of it, which he must do on the first opportunity.

105. Caution on crossing Line.—No pilot-man in search of a train, past-due, must cross on to the main line until the danger signals are turned on, and then only when he has ascertained from the police-man on duty that he can do so with safety, to prevent collision in case of the train coming up at the moment of crossing.

106. Orders Countersigned.—When any pilot engine is sent out, on its arrival at the next pilot station the engine-man must show his order to the

foreman on duty at such station, and either obtain a fresh order, or have the old one countersigned by the foreman, before he proceeds further.

107. Speed.—Any engine-man proceeding along the line with an order must be careful to proceed at the same average rate of speed as the passenger trains, and on no account to run his engine at a higher speed at any part of his journey, unless otherwise specially instructed in his order.

108. Assistance.—If both lines are obstructed, he must assist in clearing the lines, and then receive instructions for his further movements from the first guard or his superior officer, being careful not to return on the wrong line until he has satisfied himself that the police have been made aware of the circumstance for the whole distance he has to go on the wrong line; and he must then proceed slowly, sounding his whistle every quarter of a mile during the whole time.

109. Responsibility.—When any engine is sent on the line with an order, without any train, or from any other cause without a guard, the engine-man will be entirely responsible for all the movements of the engine; and in addition to the precautions and rules ordinarily applying to him, he must, in the event of being compelled to stop, send back the fire-man to make all the usual signals, as already described for stoppages. The fire-man must not return until he meets and is relieved by the next police-man. If the engine-man is obliged to cross over to the other line, and is some distance from a crossing, he must always move forward on the proper line to the next crossing, and never return on the wrong line.

110. Declaration, to be signed by each man on his appointment.

I, the undersigned, being appointed
in the service of the Company, do hereby bind myself to obey the foregoing Rules and Regulations, as well as such additional rules or instructions as I may receive from time to time. And I hereby declare that the foregoing Rules and Regulations have all been carefully read over to me and fully explained, and that I clearly understand them; that I have received a copy of them, and that I am responsible for any injury to persons or property occasioned by my neglecting the same.

(Signed)

Witness to Signature,